Planet Earth

Leonie Pratt

Designed by Zöe Wray

Illustrated by Andy Tudor

Additional illustrations by Tim Haggerty

Planet Earth consultant: Dr. Gillian Foulger, Department of Earth Sciences, Durham University
Reading consultant: Alison Kelly, Roehampton University

Contents

- 3 A place in space
- 4 All about Earth
- 6 Moving Earth
- 8 Mountain high
- 10 Volcanoes
- 12 Rock around the world
- 14 Running rivers
- 16 Wearing away
- 18 Underground caves
- 20 Cold as ice
- 22 Coast
- 24 Deep water
- 26 Dusty deserts
- 28 Extraordinary Earth
- 30 Glossary of Earth words
- 31 Websites to visit
- 32 Index

Neptune

Uranus

Saturn

A place in space

You live on Earth, one of the eight planets that move around the Sun.

Scientists think that Earth is the only planet with anything living on it.

All about Earth

Things can live on Earth because it has the right mixture of heat, air and water.

The Sun keeps the planet warm.

Living things – people, animals and plants – all need air to breathe.

Over half of the Earth is covered in water. All living things need water to stay alive.

The middle of the planet is called the core. It is very, very hot here.

Around the core is the mantle. The rocks here are so hot that they are slightly squishy.

The mantle is covered with a thin layer of solid rock, called the crust.

Moving Earth

The Earth's crust is made up of pieces.

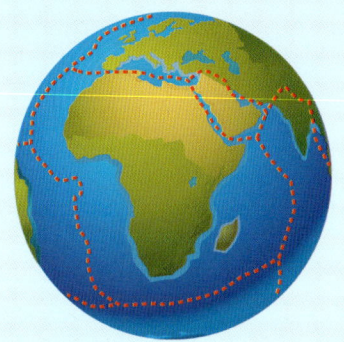 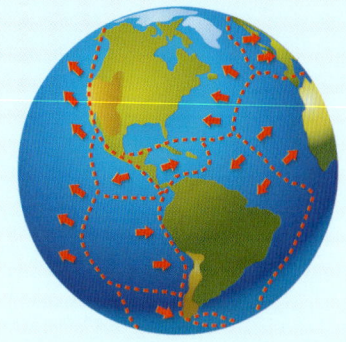

The pieces of crust slot together around the Earth.

Very very slowly, the pieces of crust move around.

The place where the pieces slide past each other is called a fault.

Pieces of crust can move smoothly along a fault, but sometimes they get jammed.

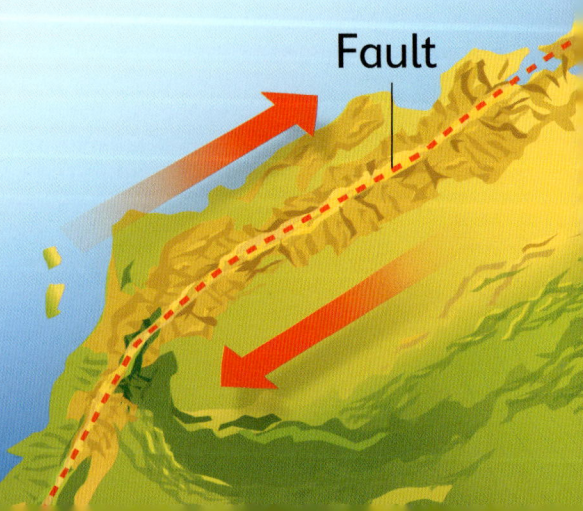

Fault

Jammed pieces can suddenly start to slip past one another. The land above trembles and cracks. This is called an earthquake.

Big earthquakes can destroy buildings and tear up roads.

There are even 'earthquakes' on the moon. These are called moonquakes.

Mountain high

As the crust moves, it slowly pushes some of the land up into tall, rocky mountains.

The mountains in this picture are the Alps, in Europe. It took millions of years for them to get this tall.

Some mountains, such as the Himalayas, are growing taller every year.

Mountain peaks stay snowy because the air is cold high up...

... so mountain goats need thick coats to keep them warm.

Volcanoes

Hot rock from the mantle can creep up through cracks in the crust and escape onto the land as a volcano.

This is a volcano erupting.

The hot rock that escapes is called lava.

Some people in Hawaii think that the goddess of fire lives in a volcano.

Volcanoes erupt in different ways.

Violent eruptions fling ash and gas high into the sky.

Sometimes volcanoes throw out lumps of hot, sticky lava.

Some volcanoes spray runny lava from a long crack in the ground.

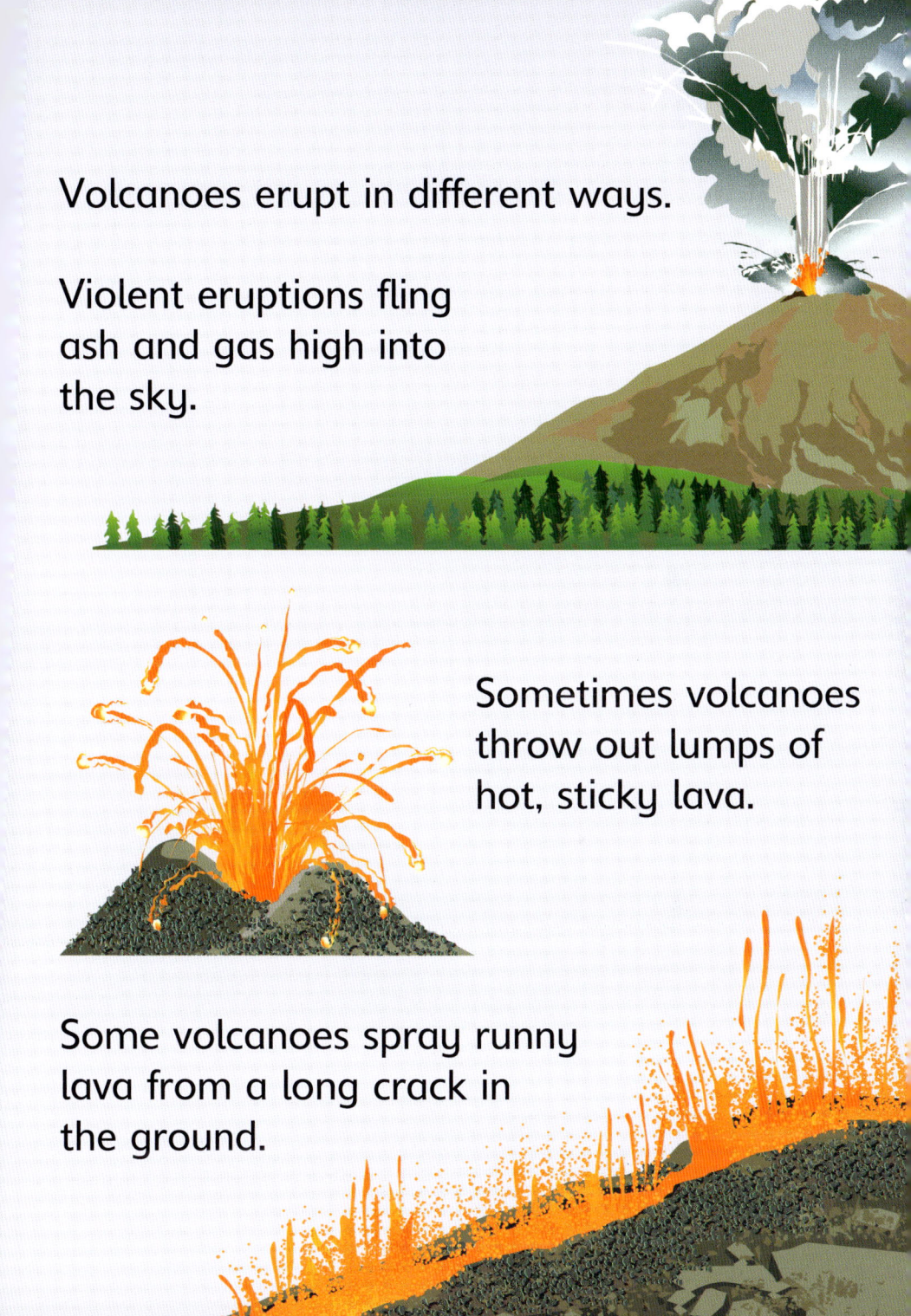

Rock around the world

All the land on Earth is made of rock. There are three main types of rock.

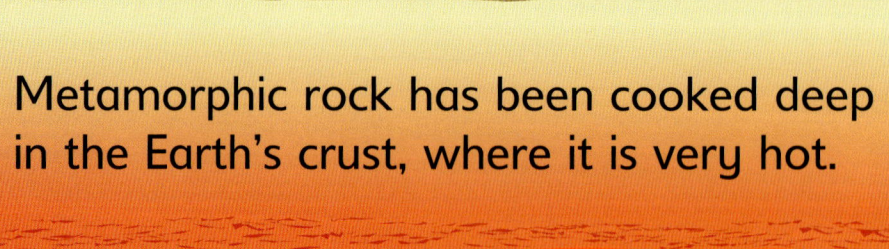

Metamorphic rock has been cooked deep in the Earth's crust, where it is very hot.

Igneous rock forms when lava from a volcano cools in the air and turns hard.

Jewels such as diamonds, rubies and emeralds are found inside rocks.

Sedimentary rock is made from layers of mud or sand pressed together.

The different layers in this rock make it look striped.

Running rivers

When it rains or snows on the mountains, the hard rock cannot soak up all the water.

Water trickles down from the high ground in a small stream.

Lots of streams join together to make a big river.

The river flows downhill, all the way to the sea.

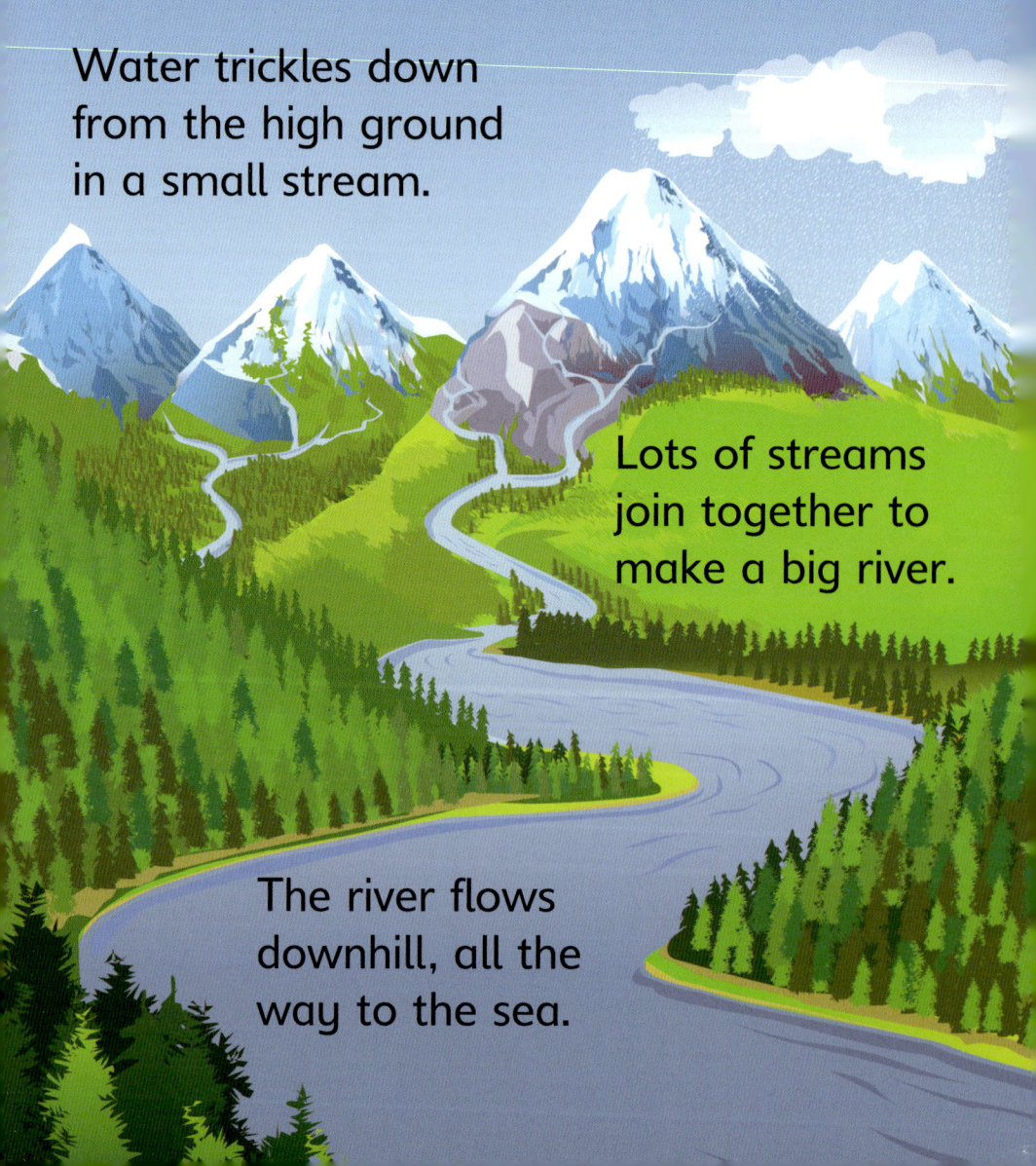

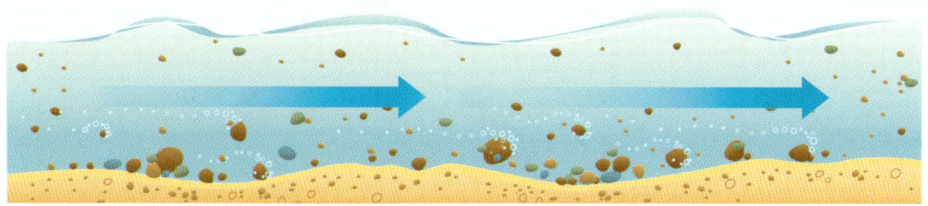

A fast-flowing river picks up lots of stones and pebbles from the riverbed.

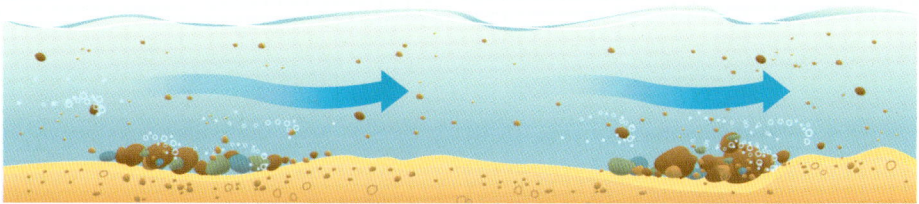

When the river slows down, it drops the heavier stones, but still carries small pebbles.

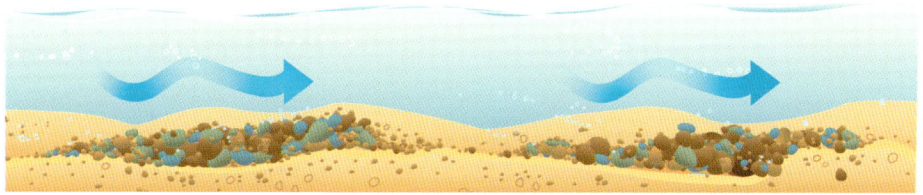

As the river gets near the sea, it slows down so much that it drops almost everything.

The Nile in Africa is the longest river in the world. It is so big it can even be seen from space.

Wearing away

Rivers change the shape of the land as they flow over it.

A river flows over the land, picking up stones and soil from the ground.

The stones bump along the riverbed, wearing away a groove in the land.

Over many years, the river carves a valley into the ground below.

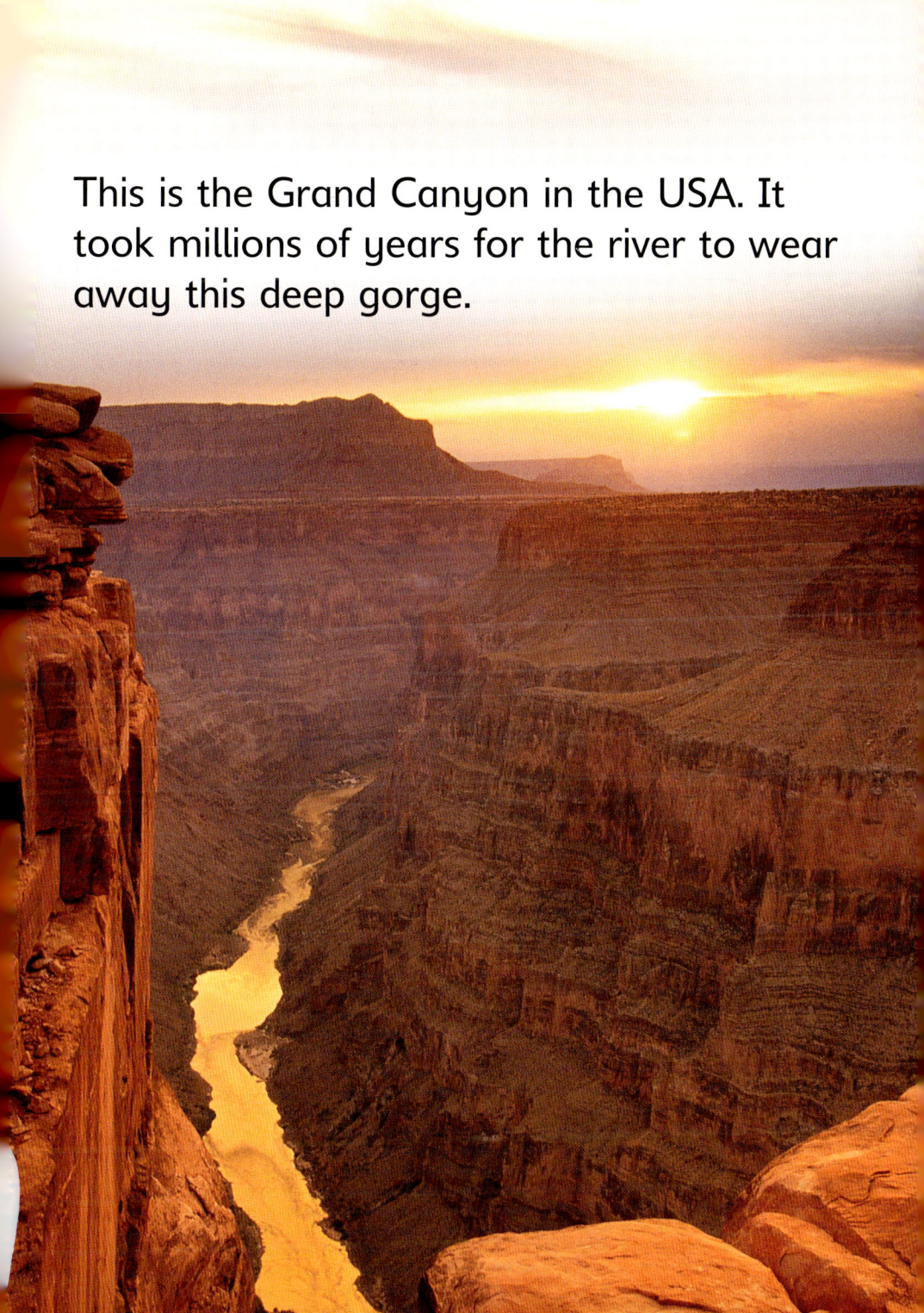

This is the Grand Canyon in the USA. It took millions of years for the river to wear away this deep gorge.

Underground caves

Not all rivers flow over the land – some flow under it. Rivers that flow underground can wear away the rock to make caves.

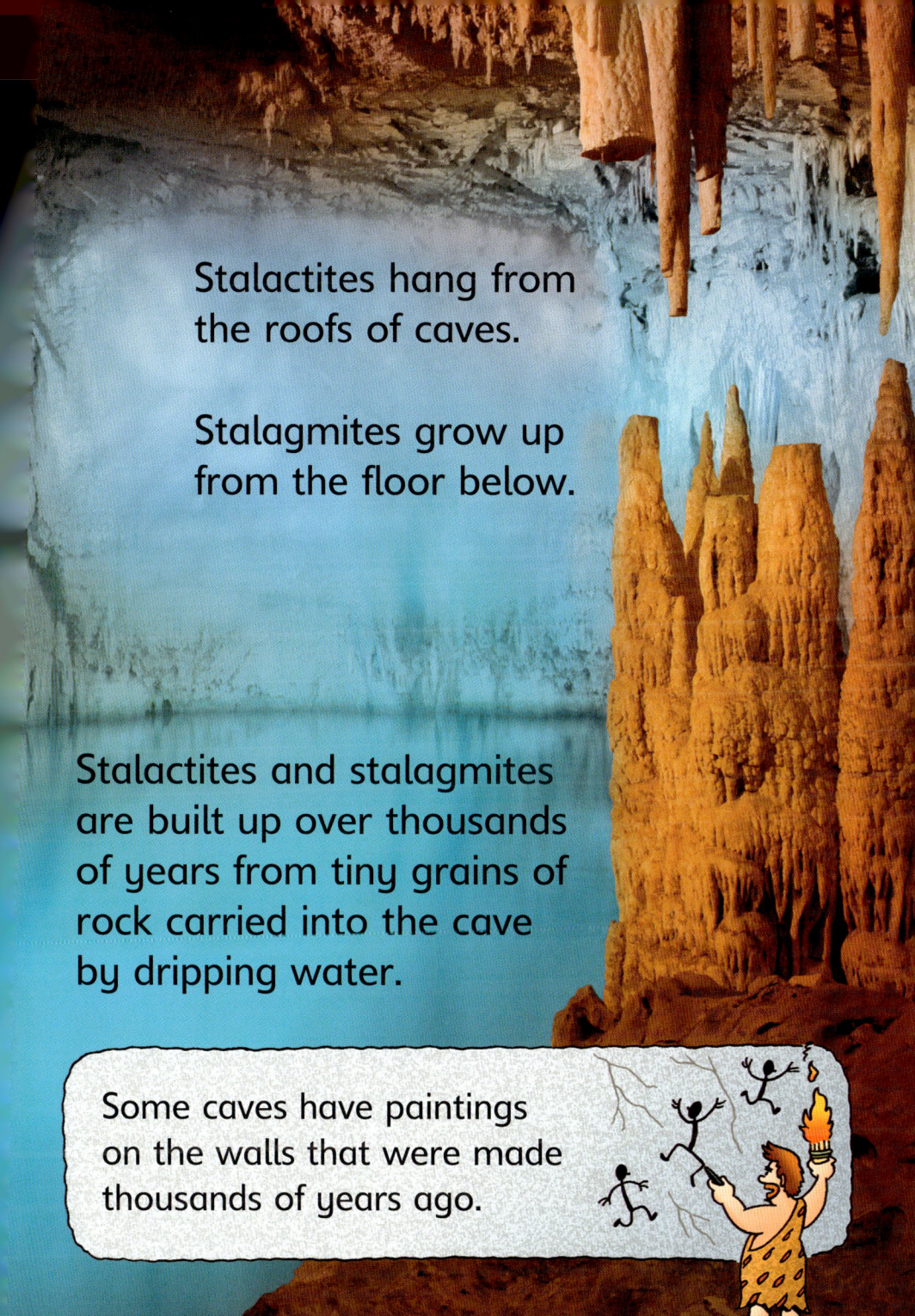

Stalactites hang from the roofs of caves.

Stalagmites grow up from the floor below.

Stalactites and stalagmites are built up over thousands of years from tiny grains of rock carried into the cave by dripping water.

Some caves have paintings on the walls that were made thousands of years ago.

Cold as ice

Some of the Earth's water stays frozen as ice for most of the year.

Antarctica is the coldest place in the world.

The land is covered with snow, and icebergs float in the sea.

1. Ice spreads from the land and floats on the sea.

2. The sea moves up and down, causing the ice to crack.

In Antarctica, it gets so cold in winter that even the sea freezes over.

3. The crack gets bigger until a block of ice breaks off.

4. The block floats away in the sea as an iceberg.

Coast

The coast is where the land meets the sea.

Sand is made from broken up shells and rocks. The sea grinds the pieces into tiny grains, then washes them ashore.

Waves crash against the sides of a cliff, wearing the rock away into an arch.

The top of the arch falls into the water, leaving a stack standing in the sea.

This is a stack.

Some beaches have black sand made from volcanic rock.

Deep water

Oceans and seas cover more than half of the Earth. Many different things live there.

Lots of creatures live near the surface, where it is light and the water is warm.

Deeper down it is darker and colder. Fewer things can live here.

Sperm whales dive deep down looking for squid to eat.

Scientists use submarines like this one to dive deep into the sea and explore the ocean floor.

Some deep sea fish glow in the dark to attract small fish – then they eat them!

Dusty deserts

Deserts are the driest places on Earth. Very little rain falls and the land is dry and dusty.

The Sahara desert in Africa is one of the biggest deserts in the world.

1. Even in the desert, water is trapped in rocks underground.

2. Over a long time, the water builds up and forms a pool.

Desert animals stay underground during the hot days and come out at night to find food.

3. Plants grow near the pool. This is called an oasis.

4. People who travel across deserts often stop at an oasis.

Extraordinary Earth

Planet Earth is an amazing place...

Greenland is the largest island in the world.

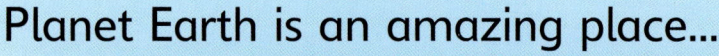

Angel Falls, Venezuela is the world's tallest waterfall. It is 979m (3,212ft) high.

Some parts of the Atacama desert in Chile, went without rain for 400 years.

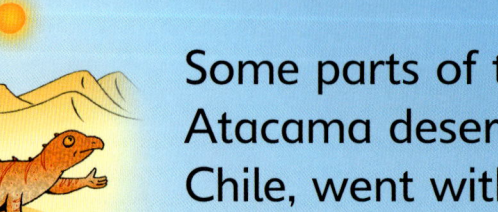

Mount Everest is the highest place in the world. It is 8,850m (29,035ft) high.

The deepest part of the Marianas Trench is about 11,000m (36,089ft) underwater.

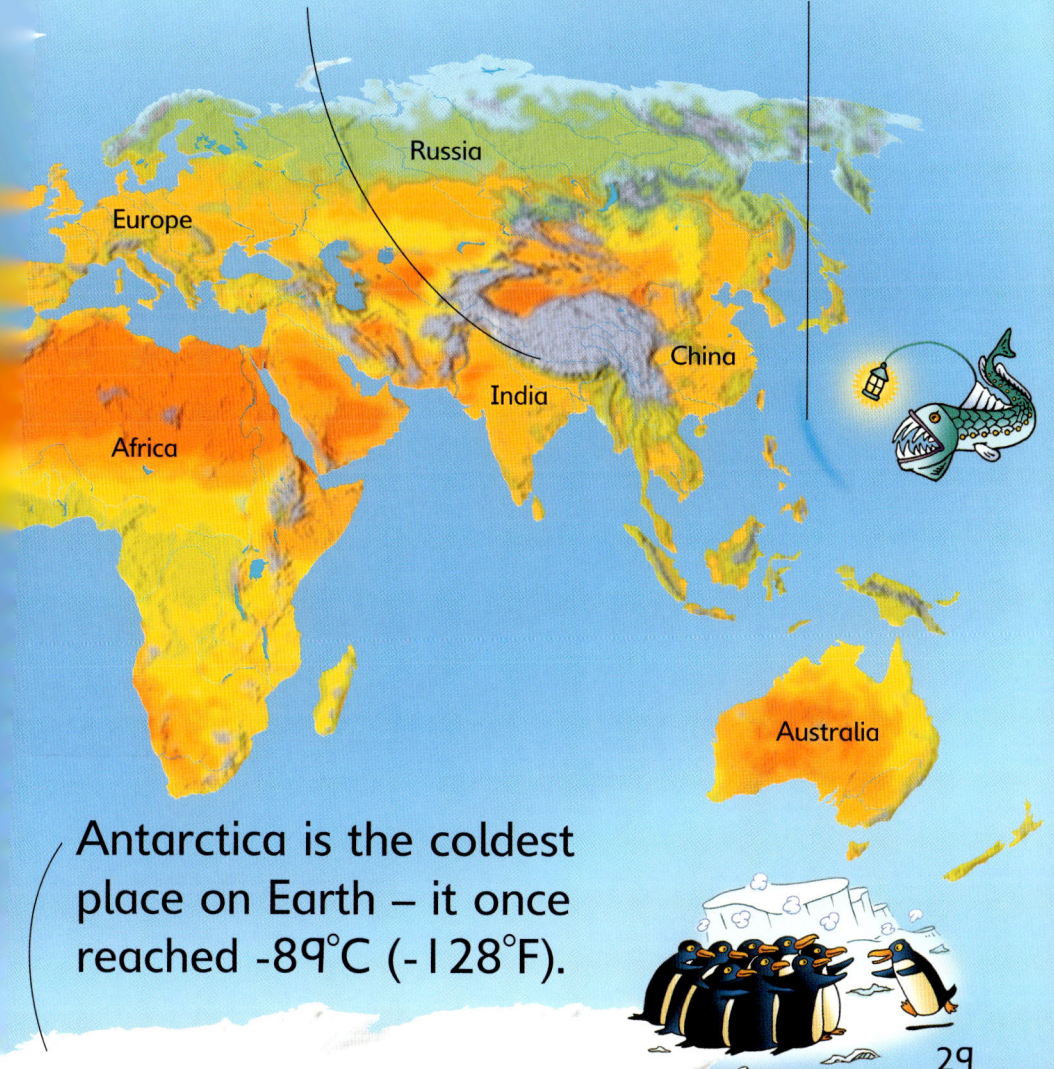

Antarctica is the coldest place on Earth – it once reached -89°C (-128°F).

Glossary of Earth words

Here are some of the words in this book you might not know. This page tells you what they mean.

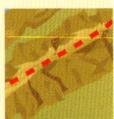 fault - the place where two pieces of crust meet.

 eruption - when hot rock escapes onto the land from a volcano.

 lava - red-hot melted rock that has erupted from a volcano.

 gorge - a deep valley, with steep sides that have been carved out by a river.

 iceberg - a big block of ice that floats in the sea. Most of it is underwater.

 desert - a place where there is very little rainfall in a year.

 oasis - a pool in a desert where water has risen up from below the ground.

Websites to visit

You can visit exciting websites to find out more about Planet Earth.

To visit these websites, go to the Usborne Quicklinks Website at **www.usborne-quicklinks.com** Read the internet safety guidelines, and then type the keywords "**beginners planet earth**".

The websites are regularly reviewed and the links in Usborne Quicklinks are updated. However, Usborne Publishing is not responsible, and does not accept liability, for the content or availability of any website other than its own. We recommend that children are supervised while on the internet.

These massive rocks are in Monument Valley, USA.

Index

Africa, 15, 26, 29
Antarctica, 20, 21, 29
caves, 18-19
coast, 22-23
core, 5
crust, 5, 6, 8, 10, 12
deserts, 26-27, 28, 30
earthquakes, 6-7
Europe, 8, 29
faults, 6, 30
icebergs, 20-21, 30
lava, 10, 11, 12, 30
mantle, 5, 10
mountains, 8-9, 14, 29
North America, 17, 28, 31
rivers, 14-15, 16, 17, 18, 30
rocks, 5, 8, 10, 12-13, 14, 18, 19, 22, 23, 26, 30
sand, 13, 22, 23
seas and oceans, 14, 15, 20, 21, 22, 23, 24-25, 29, 30
Sun, 3, 4
volcanoes, 10-11, 12, 23, 30
wearing away, 16-17, 18, 23

Acknowledgements

Additional design by Helen Wood and Erica Harrison
Map illustration page 28-29 by Craig Asquith, European Map Graphics Ltd.
Photographic manipulation by John Russell

Photo credits

The publishers are grateful to the following for permission to reproduce material:
© BRUCE COLEMAN INC./Alamy cover; © Digital Vision 31; © Frans Lemmens/zefa/Corbis 26-27;
© Gabe Palmer/CORBIS 1; © Henry Westheim Photography/Alamy 7; © image broker/Alamy 18-19;
© Jim Sugar/Corbis 10; © Joel Simon/Digital Vision 20-21; © Joseph Sohm/Visions of America/Corbis 13; © Marc Garanger/CORBIS 8-9; © Michael Howard/Alamy 22-23; © NASA 2-3, 5; © Photo by Rod Catanach, Woods Hole Oceanographic Institution 25; © Ron Watts/CORBIS 17

Every effort has been made to trace and acknowledge ownership of copyright. If any rights have been omitted, the publishers offer to rectify this in any subsequent editions following notification.

First published in 2007 by Usborne Publishing Ltd., Usborne House, 83-85 Saffron Hill, London EC1N 8RT, England. www.usborne.com Copyright © 2007 Usborne Publishing Ltd. The name Usborne and the devices ⓠ⊕ are Trade Marks of Usborne Publishing Ltd. All rights reserved. No part of this publication may be reproduced, stored in a retrieval system, or transmitted in any form or by any means, electronic, mechanical, photocopying, recording or otherwise without the prior permission of the publisher.
First published in America 2007. U.E. Printed in China.